YOUR KNOWLEDGE HAS VALUE

- We will publish your bachelor's and
  master's thesis, essays and papers

- Your own eBook and book -
  sold worldwide in all relevant shops

- Earn money with each sale

Upload your text at www.GRIN.com
and publish for free

Christian Momberger

Climate Atlas of Eurasian and North American Arctic regions

GRIN Verlag

Bibliografische Information der Deutschen Nationalbibliothek:

Die Deutsche Bibliothek verzeichnet diese Publikation in der Deutschen National-
bibliografie; detaillierte bibliografische Daten sind im Internet über http://dnb.d-
nb.de/ abrufbar.

Imprint:

Copyright © 2001 GRIN Verlag GmbH
Druck und Bindung: Books on Demand GmbH, Norderstedt Germany
ISBN: 978-3-640-13535-6

This book at GRIN:

http://www.grin.com/en/e-book/113747/climate-atlas-of-eurasian-and-north-ame-
rican-arctic-regions

Arctic Studies Program 2001-2002
Ecological and Historical Biogeography of Arctic and Alpine Regions

Autumn Term 2001

Climate Atlas

of Eurasian and North American Arctic regions

Author: Christian Momberger, Gießen, Germany

Contents:

1. Introduction

The following climate atlas will show the climatic differences between selected arctic regions from Eurasia, or strictly speaking Russia, and North America. The main emphasis will thereby be put on the difference between oceanic (maritime) and continental climate. It will also show the spread of permafrost and the localisation of the Arctic tree line, which both depends on the existing climate.

But what are the differences between oceanic and continental climate? How could the climates be described?
Oceanic climates can be found in coastal regions or a bit more inland on the western side of Eurasia or eastern parts of North America. They characterise themselves through a high rate of total precipitation - wet climate - and a low annually temperature amplitude - the range between the temperature of the coldest month, normally January on northern hemisphere, and the temperature of the warmest month, normally July. An example for such a climate is shown in *Figure 1a*. Here the annually amplitude is "$\Delta t = 29\ °C$" and the total amount of precipitation is "*917 mm*". As you can see, you will always find enough precipitation, so there will be no dry-period. A dry-period is defined as following:

$$precipitation\ [mm] < temperature * 2\ [°C]$$

As you move inland, away from the western coasts of Eurasia or eastern coasts of North America, the climate will become more continental. This means, that the total precipitation will decrease and the temperature range will increase. A model for an extreme continental climate is given in *Figure 1b*. As you can see the temperature amplitude is "$\Delta t = 61\ °C$" - more than two times of that

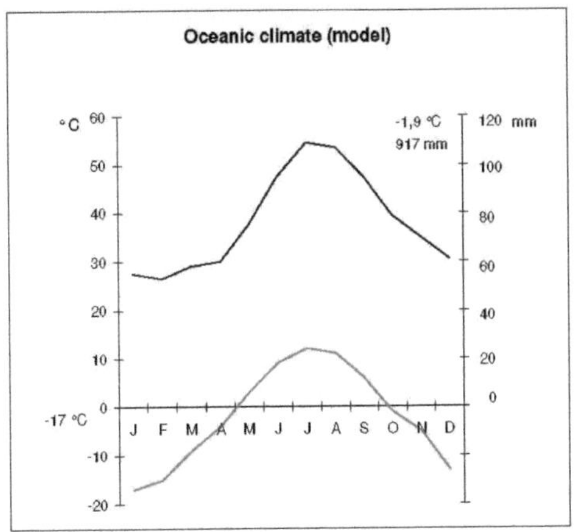

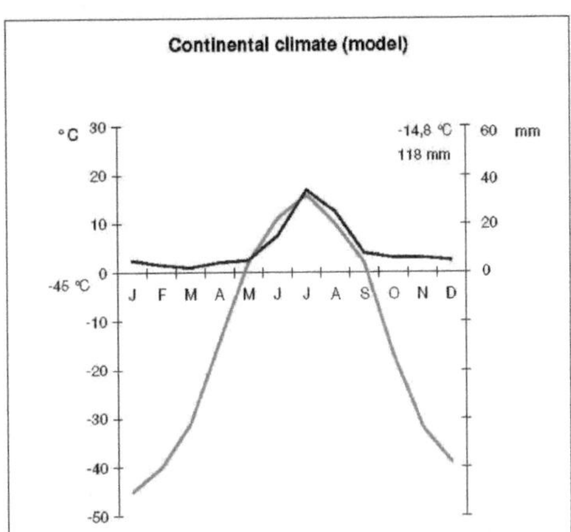

shown in the model for the oceanic climate *(Figure 1a)* - and the total amount of precipitation is only *"118 mm"*. In this example you can also find a dry-period in early summer, where you have little precipitation and a strong rising temperature. Continental climate stations will show a much lower temperature of the coldest month than oceanic stations (here: -45 °C to -17 C).
In areas with continental climates you will have a less depth of snow cover than in oceanic regions for the reason of the lower precipitation.

The arctic can be divided into two vegetation zones, the Taiga, consisting of coniferous forest and the Tundra, normally treeless and covered with lichen and moss. The border between these two zones, which cannot easily marked, will be the Artic tree line.
The lower precipitation, the less snow cover, the colder temperature during winter and the wider annual range, will influence the localisation of the Arctic tree line and also the existence of permafrost. You can distinguish four types of permafrost by their extension:

- *Sporadic permafrost:*   *< 1% land cover*
- *Isolated permafrost:*   *1% < 10% land cover*
- *Discontinuous permafrost:* *10% < 50% land cover*
- *Continuous permafrost:*   *> 50% land cover*

The Arctic tree line and permafrost will be short discussed in chapter 3.

In the Appendix, the climate-diagrams for each below given station and equivalent maps of the chosen sections and for the locating of the tree line and the permafrost can be found.

2. Arctic climate stations

In this chapter the climatic variation within the arctic will be shown. Therefore four different south-north cross-sections are chosen - each two from Eurasia (Russia) and Northern America (Canada). As a fifth example the most continental climate station in North America, Fort Yukon (Alaska), is shown. For each station, if available, will be given:

- the region and the country within the station is located
- the altitude above sees level

- the mean monthly and annual temperature
- the monthly precipitation and the total amount of it
- the temperature range between the coldest and the warmest month
- the extreme minimum temperature
- the vegetation zone within the station is localised
- the type of permafrost that can be found

The following sections are shown from south to north - the most southern station of each section will be first. In every section, if possible, will be a station, situated close to the tree line and/or the southern limit of Discontinuous permafrost.

2.1. Siberian Section

The first section, which is chosen here, is a section trough Siberia, starting in central Manchuria (China) and moving north up to the Arctic Ocean. The last station, Bulun, is located a bit south of the mouth of the river Lena. This section gives an example of a hyper-continental arctic climate: The temperature range is extremely high - between 41 °C (Harbin) and 63 °C (Verkhoyansk) - and the total precipitation is, excluded the southern cities Harbin and Blagoveshchensk, below 200 mm. In Yakutsk and Verkhoyansk you can find a dry-period in early summer.

Table 1 shows the climate-table for the northern Chinese city of Harbin. It is located within the Asian step, a dry, short grass covered zone of Inner-Asia. Here will you will find isolated permafrost for the reason that the temperatures in winter are below freezing while having only less snow cover corresponding to the little precipitation. In summer the temperature is about 20 °C, so that the range is 41 °C. The most precipitation will be during summer and for that there is no dry-period (see *Figure 2* in the Appendix).

Table 1: Harbin (160 m), Manchuria (China)

Month	J	F	M	A	M	J	J	A	S	O	N	D	Year
Mean Monthly Temperature (°C)	-18	-14	-4	6	14	19	23	21	14	4	-6	-16	3,6
Mean Monthly Precipitation (mm)	5	5	10	23	43	94	112	104	46	33	8	5	488

Temperature range	41 °C	Vegetation zone	Asian step
Extreme minimum	-37,8 °C	Permafrost type	Isolated

The next station within the Siberian section is Blagoveshchensk; his data are given in *Table 2*. This city can be located on the river Amur within the vegetation zone of the Taiga, but you will also recognise trees growing in the mixed-forest of mid-latitudes. Here you have already discontinuous permafrost, which means that a larger area is covered by it than in the southern Harbun. Therefore the reason for the existence of permafrost will be the same. The most precipitation will be in summer and the temperature range is 44 °C. This station is also shown in *Figure 3* in the Appendix.

Table 2: Blagoveshchensk (142 m), Siberia (Russia)

Month	J	F	M	A	M	J	J	A	S	O	N	D	Year
Mean Monthly Temperature (°C)	-22	-18	-9	2	11	17	22	21	12	3	-12	-21	0,5
Mean Monthly Precipitation (mm)	2	3	8	23	41	84	112	114	69	18	8	3	485

Temperature range	44 °C	Vegetation zone	Taiga/mixed forest
Extreme minimum		Permafrost type	Discontinuous

The third Siberian station is Yakutsk *(Table 3)*. It is located in central eastern Siberia within the Taiga vegetation zone. Here the climate is much more continental than in the two above shown southern station of this section. The total amount of precipitation will be only 187 mm and the temperature range is 61 °C. The mean temperature will be much colder during the year and so the permafrost extension can be bigger - more than 50% of the area are covered (continuous permafrost). But one can still find trees for the reason, that the mean temperature is (still) above +10 °C for more than 30 days (one month). In early summer (May - July) there is a dry-period, where you have not enough precipitation (see *Figure 4* in the Appendix). The precipitation maximum can be recognised in August one month after the climate maximum.

Table 3: Yakutsk (102m), Siberia (Russia)

Month	J	F	M	A	M	J	J	A	S	O	N	D	Year
Mean Monthly Temperature (°C)	-43	-35	-21	-9	7	15	18	14	6	-8	-28	-41	-10,4
Mean Monthly Precipitation (mm)	7	6	5	7	12	24	32	39	20	14	12	9	187

Temperature range	61 °C		Vegetation zone	Taiga
Extreme minimum	-57,0 °C		Permafrost type	Continuous

Table 4 shows the data for the Siberian climate station Verkhoyansk (northern Siberia), located also in the Taiga vegetation zone. The temperature will be above +10 °C in summer (June-August), so trees can grow, and extremely cold in winter. The rage is 65 °C and the extreme minimum temperature (-70 °C) is the coldest of all here chosen stations. The total amount of precipitation is 102 mm. By that in June you will have a dry-period like in Yakutsk (see *Figure 5* in the Appendix). The permafrost type, than can be found is continuous permafrost.

Table 4: Verkhoyansk (100 m), Siberia (Russia)

Month	J	F	M	A	M	J	J	A	S	O	N	D	Year
Mean Monthly Temperature (°C)	-50	-45	-30	-13	2	12	15	11	2	-21	-37	-46	-16,7
Mean Monthly Precipitation (mm)	5	2	1	2	5	13	31	23	5	5	5	5	102

Temperature range	65 °C		Vegetation zone	Taiga
Extreme minimum	-70,0 °C		Permafrost type	Continuous

The last and most northern station of this section is Bulun *(Table 5)*. It is already located north of the Arctic tree line in the Tundra-zone. Even the mean temperature in July is above +10 °C, this period may to short for tree-growth and the precipitation may to little. Unfortunately the monthly amounts of precipitation are not available but the total sum is below 100 mm. The temperature range is smaller than in Yakutsk or Verkhoyansk, because of its location not to far from the Arctic Ocean, but still 53 °C. For this station see also *Figure 6* in the Appendix-

Table 5: Bulun (35 m), Siberia (Russia)

Month	J	F	M	A	M	J	J	A	S	O	N	D	Year
Mean Monthly Temperature (°C)	-41	-36	-26	-16	-4	7	12	9	2	-23	-30	-36	-15,2
Mean Monthly Precipitation (mm)													< 100

Temperature range	53 °C		Vegetation zone	Tundra
Extreme minimum			Permafrost type	Continuous

As you can see, moving north through Siberia, the climate will get more and more continental, the temperature range increases while the amount of precipitation decreases at the same time. In this section the Arctic tree line can be located somewhere between the stations of Verkhoyansk and Bulun (mean July temperature is 15 °C respectively 12 °C). The discontinuous permafrost reaches as far south as somewhere between the stations of Blagoveshchensk and Harbin (mean annual temperature is 0,5 °C respectively 3,6 °C). Considering these two locations, you can recognise a distance of about 1.000 km between the Arctic tree line and the southern limit of discontinuous permafrost.

2.2. Komi-Nenets Section

The second chosen section is a section of three stations in the Komi-Nenets area, laying nearly close together and located west of the northern Ural Mountain in Russia. This section gives an example for a sub-continental climate with an intermediate temperature range (about 33 °C) and total amount of precipitation (around 500 mm). The northern station Khoseda-Khard is located at the Arctic tree line.

The most southern station of this short section is Pechora, shown in *Table 6*. It is located in the northern parts of the Taiga vegetation zone. The mean temperature in summer will be three months above +10 °C and in winter not below -20 °C, the range is 35,5 °C. The precipitation spreads all over the year with a little peak in autumn. Here you can find isolated permafrost. This station is also shown in *Figure 7* in the Appendix.

Table 6: Pechora, Komi (Russia)

Month	J	F	M	A	M	J	J	A	S	O	N	D	Year
Mean Monthly Temperature (°C)	-19,5	-17,7	-11,6	-3,4	3,4	11,1	16,0	12,3	6,1	-2,5	-10,6	-15,6	-2,7
Mean Monthly Precipitation (mm)	37	29	30	31	41	47	53	70	72	59	48	39	556

Temperature range	35,5 °C	Vegetation zone	Taiga (northern)
Extreme minimum		Permafrost type	Isolated permafrost

The data for the second station in this section, Ust'-Usa, are given in *Table 7*. It is also situated in the northern parts of the Taiga-zone and has a short summer with temperatures above +10 °C. The precipitation spreads throughout the year with a peak in August and September – the total amount is 495 mm. This station can be located in the area of the southern limit of discontinuous permafrost. For this station see also *Figure 8* in the Appendix.

Table 7: Ust'-Usa, Komi (Russia)

Month	J	F	M	A	M	J	J	A	S	O	N	D	Year
Mean Monthly Temperature (°C)	-18,4	-17,6	-12,9	-4,3	1,4	9,5	14,1	11,6	5,7	-2,1	-9,6	-15,6	-3,2
Mean Monthly Precipitation (mm)	30	22	24	27	36	46	53	63	66	53	41	34	495

Temperature range	32,5 °C	Vegetation zone	Taiga (northern)
Extreme minimum		Permafrost type	Discontinuous (southern limit)

Table 8: Khoseda-Khard, Komi (Russia)

Month	J	F	M	A	M	J	J	A	S	O	N	D	Year
Mean Monthly Temperature (°C)	-19,6	-19,5	-15,8	-7,6	-1,1	7,4	12,6	10,1	4,8	-3,5	-11,2	-16,7	-5,0
Mean Monthly Precipitation (mm)	23	20	20	21	32	43	52	61	62	47	32	23	436

Temperature range	32,2 °C	Vegetation zone	Arctic tree line
Extreme minimum		Permafrost type	Discontinuous

The third station Khoseda-Khard (Table 8), located a bit north of the Arctic Circle, is situated in the area of the Arctic tree line. Here the temperature range is 32,2 °C and the total precipitation 436 mm (spread all over the year). It lies within the zone of discontinuous permafrost. A climate-diagram for this station is drawn in *Figure 9* in the Appendix.

In this section the distance between the Arctic tree line, located at Khoseda-Khard (mean July temperature 12,6 °C), and the southern extension of discontinuous permafrost at Ust'-Usa (mean annual temperature -3,2 °C) is much shorter than within the Siberian section. For the fact, that the stations are relatively close together, they are not so much different from each other.

2.3. Quebec Section

One of the two North-American sections chosen in here is going through central Quebec, an Eastern-Canadian province. The section, consisting of three climate-stations, is moving from southeast Quebec to western Quebec, at the coast of the Hudson Bay. It is a very good example for an oceanic climate in the Arctic. In this region you will find high precipitation (total amount between 962 mm and 631 mm) and a temperature range of around 35 °C. You won't find any dry-period.
The climate-data for the Quebec section are given for the latest normal-period 1961-1990.

The first station of this section is Manicouagan (*Table 9*), located east of the Gulf of St. Lawrence. It lies within the Taiga vegetation zone and has the highest precipitation of all here chosen arctic stations. You will find enough precipitation during the whole year but more in summer than in winter. Here you have isolated permafrost, which means that only little areas are affected. This fact is the result of the high rate of precipitation, even in winter, and by that the much higher snow-cover, so that the cold winter-air cannot penetrate the ground. Sometimes, the temperature gets, even it is an oceanic station, very cold, but only for one or a few day. The extreme minimum temperature is -51,1 °C. *Figure 10* in the Appendix shows the climate-diagram for this station

Table 9:Manicouagan (406 m), Quebec (Canada)

Month	J	F	M	A	M	J	J	A	S	O	N	D	Year
Mean Monthly Temperature (°C)	-21	-19	-13	-3	4	12	14	13	8	1	-7	-17	-2,3
Mean Monthly Precipitation (mm)	66	51	55	47	64	112	119	106	110	82	68	82	962

Temperature range	35 °C	Vegetation zone	Taiga (middle)
Extreme minimum	-51,1 °C	Permafrost type	Isolated

Table 10: Nitchequon (536 m), Quebec (Canada)

Month	J	F	M	A	M	J	J	A	S	O	N	D	Year
Mean Monthly Temperature (°C)	-23	-22	-16	-5	2	9	13	11	7	-1	-6	-18	-4,1
Mean Monthly Precipitation (mm)	40	30	36	38	58	86	98	102	96	87	66	46	783

Temperature range	36 °C	Vegetation zone	Tundra (southern)
Extreme minimum	-49,4 °C	Permafrost type	Isolated

Table 10 shows the data for the station Nitchequon, central Quebec. It lies within the southern zone of Tundra-vegetation, but you will although find some trees, related to the fact, that there a more than 30 days with a temperature above +10 °C. The precipitation is a bit lower (total amount 783 mm) than in Manicouagan, but there will still enough precipitation in winter to form a good snow-cover that prevents most parts of the ground of permafrost. So one will find only isolated

permafrost. The temperature range is 36 °C. The climate-diagram is given in *Figure 11* in the Appendix.

The last and most northern station of the Quebec-Section is Poste-de-la-Baleine, shown in *Table 11*. It is situated at the coast of the Hudson Bay in the area of the Arctic tree line and the southern limit of discontinuous permafrost. As you can see, the precipitation is lower than at the two other climate-stations but still 631 mm during the year. The temperature range is equal compared to the other stations. For this station see also *Figure 12* in the appendix.

Table 11: Poste-de-la-Baleine (16m), Quebec (Canada)

Month	J	F	M	A	M	J	J	A	S	O	N	D	Year
Mean Monthly Temperature (°C)	-23	-23	-16	-6	2	5	11	10	6	3	-4	-17	-4,3
Mean Monthly Precipitation (mm)	26	24	22	27	41	53	86	96	89	63	61	43	631

Temperature range	34 °C		Vegetation zone	Arctic Tree line
Extreme minimum	-49,4 °C		Permafrost type	Discontinuous (southern limit)

Comparing the three climate-stations of the Quebec section, one will find only little differences. In this region, with the most oceanic climate of the chosen cross-sections, the Arctic tree line and the southern limit of permafrost are located nearly close together in the area around Poste-de-la-Baleine (mean annual temperature -4,3 °C, mean July temperature 10 °C) *(see Table 11)*. As you can see, while moving westwards, away from the Atlantic Ocean, the total amount of precipitation will decrease.

2.4. Central-Canada Section

The last chosen section is the Central-Canada section, going from Medicine Hat in the Canadian Prairie close to the border to the United States all the way north to Alert nearly at the top of Ellesmere Island, "the north cape of Canada". It is the longest section, including ten climate-stations. Here you will have a continental climate and find differences between the stations within this section. The Temperature range varies form 32,5 °C in Medicine Hat up to 44,0 °C in Baker Lake (southern Northwest Territories) and the total amount of precipitation is between 489,4 mm at La Ronge (central Saskatchewan) and 131,4 mm at Resolute Airport (central Northwest Territories). The data for the Canadian climate-stations are given for the latest normal period 1961-1990.

The most southern station of this section is Medicine Hat *(Table 12)* in Southeast Alberta, one of the so-called Canadian "Prairie-provinces". It is situated within the North American Prairie, a vegetation zone consisting of wide-open long grass covered plains. Here you won't find any permafrost - the station lies at to southern latitude. Although the mean annual temperature is +5,1 °C, some winter day might be very cold - the extreme minimum is -46,1 °C. In August, you can recognise a small dry-period (see *Figure 13* in the Appendix). The most precipitation will be recognised in early summer.

Table 12: Medicine Hat (717 m), Alberta (Canada)

Month	J	F	M	A	M	J	J	A	S	O	N	D	Year
Mean Monthly Temperature (°C)	-12,6	-7,7	-2,8	5,6	12,3	16,6	19,9	18,9	13,2	7,4	-1,6	-7,6	5,1
Mean Monthly Precipitation (mm)	22,7	16,6	18,5	30,2	40,1	63,5	40,4	36,4	32,4	16,2	14,6	16,3	347,9

Temperature range	32,5 °C		Vegetation zone	Prairie
Extreme minimum	-46,1 °C		Permafrost type	No permafrost

The next northern station is Dorintosh in western Saskatchewan, also one of the so-called "Prairie-Provinces" of Canada. The data therefore are given in *Table 13*. It is characterised by a temperature range of 37,2 °C and a total amount of precipitation of 439,1 mm. It is wetter than in Medicine Hat and you won't find any dry-period. The station is situated in the southern party of the Taiga vegetation zone and the area of the southern extension of isolated permafrost. The corresponding climate-diagram is *Figure 14* in the Appendix.

Table 13: Dorintosh (503 m), Saskatchewan (Canada)

Month	J	F	M	A	M	J	J	A	S	O	N	D	Year
Mean Monthly Temperature (°C)	-20,7	-15,7	-9,0	2,2	9,9	14,2	16,5	14,9	9,2	3,7	-7,2	-16,3	0,1
Mean Monthly Precipitation (mm)	15,6	14,1	23,6	21,5	33,6	65,4	73,9	77,6	53,4	18,1	20,9	21,4	439,1

Temperature range	37,2 °C	Vegetation zone	Taiga (southern)
Extreme minimum	-50,6 °C	Permafrost type	Isolated (southern limit)

In central Saskatchewan located is the climate-station of La Ronge (*Table 14*). It lies within the central parts of the Taiga and has a total amount of precipitation of 489,4 mm - the highest precipitation rate within the Central-Canadian section. The mean annual temperature is just below freezing (-0,5 °C). In the area of La Ronge, you will find the southern limit of discontinuous permafrost. For this station see also *Figure 15* in the Appendix.

Table 14:La Ronge (375 m), Saskatchewan (Canada)

Month	J	F	M	A	M	J	J	A	S	O	N	D	Year
Mean Monthly Temperature (°C)	-20,9	-16,9	-9,5	1,1	8,7	14,2	16,9	15,4	9,4	2,6	-8,9	-18,0	-0,5
Mean Monthly Precipitation (mm)	18,9	15,1	18,5	27,1	40,1	79,4	84,3	60,3	59,3	35,4	29,1	21,8	489,4

Temperature range	37,8 °C	Vegetation zone	Taiga (middle)
Extreme minimum	-48,3 °C	Permafrost type	Discontinuous (southern limit)

Table 15 shows the climate-table for station of Cree Lake. This station is situated in northern Saskatchewan within the central parts of the Taiga-zone. Here you will have discontinuous permafrost and a temperature range of 40,7 °C. The mean monthly temperatures is above +10 °C from June until August. The total amount of precipitation is 413,8 mm, mostly available in summer. A climate-diagram for this station is drawn in *Figure 16* in the Appendix.

Table 15: Cree Lake (497 m), Saskatchewan (Canada)

Month	J	F	M	A	M	J	J	A	S	O	N	D	Year
Mean Monthly Temperature (°C)	-25,1	-19,7	-13,6	-1,8	6,1	12,9	15,6	14,4	7,9	1,6	-10,4	-20,1	-2,7
Mean Monthly Precipitation (mm)	14,8	13,4	15,6	21,7	25,8	50,8	79,1	60,5	57,3	30,1	20,9	23,8	413,8

Temperature range	40,7 °C	Vegetation zone	Taiga (middle)
Extreme minimum	-47,8 °C	Permafrost type	Discontinuous

Table 16: Stony Rapids (213 m), Saskatchewan (Canada)

Month	J	F	M	A	M	J	J	A	S	O	N	D	Year
Mean Monthly Temperature (°C)	-28,4	-23,1	-16,4	-3,4	5,7	12,2	15,3	13,6	6,7	0,1	-13,1	-22,6	-4,5
Mean Monthly Precipitation (mm)	16,5	13,5	17,9	19,9	22,1	46,7	61,4	59,5	42,7	28,3	25,6	22,6	376,7

Temperature range	43,7 °C	Vegetation zone	Taiga (northern)
Extreme minimum	-50,6 °C	Permafrost type	Discontinuous

Also located in northern Saskatchewan is the climate-station Stony Rapids *(Table 16)*. Here you are reaching the northern parts of the Taiga. The permafrost-type that can be found at this station is discontinuous permafrost. The climate at this station is bit dryer than at Cree Lake - the total amount of precipitation is 376,7 °C. In winter you will only have little precipitation (see *Figure 17* in the Appendix). Going northwards the temperature range is increasing, here it is already 43,7 °C.

The climate-station Ennadai Lake, shown in Table 17, is located in the most southern parts of the Canadian Northwest-Territories and west of the Hudson Bay. At this station you are already reaching the Arctic tree line and the southern extension of Continuous permafrost. The mean annual temperature is -9,3 °C and the range 43,9 °C. The precipitation is less than 300 mm and in February only 6,4 mm. This is the reason, why the snow cover is less and the cold winter air can penetrate into the ground, so that you will get continuous permafrost and bad conditions for the tree growth. The most precipitation at this station will be available in summer (see *Figure 18* in the Appendix).

Table 17: Ennadai Lake (325 m), Northwest Territories (Canada)

Month	J	F	M	A	M	J	J	A	S	O	N	D	Year
Mean Monthly Temperature (°C)	-30,9	-29,1	-23,8	-13,1	-2,7	7,2	13,0	11,5	3,9	-5,0	-17,3	-25,8	-9,3
Mean Monthly Precipitation (mm)	10,3	6,4	11,7	15,3	19,6	30,7	51,9	41,7	44,7	33,2	16,4	12,6	294,5

Temperature range	43,9 °C		Vegetation zone	Arctic tree line
Extreme minimum	-50,3 °C		Permafrost type	Continuous (southern limit)

The sevenths station in the Central-Canadian section from south to north is Baker Lake, which is laying in the southern parts of the Northwest Territories and west of the northern parts of the Hudson Bay. His data are given in *Table 18*. The vegetation zone of this station is the Tundra or strictly speaking it's southern parts. The permafrost type is continuous permafrost and the total amount of Precipitation is 234,6 mm. In winter you will have very little precipitation and so less snow-cover. Even the mean July temperature is bit above +10 °C, you will find an abundance of forests. This may related to the fact, that there a less than thirty day with a temperature above the +10 °C. The climate-diagram for this station is drawn in *Figure 19* in the Appendix.

Table 18: Baker Lake (12 m), Northwest Territories (Canada)

Month	J	F	M	A	M	J	J	A	S	O	N	D	Year
Mean Monthly Temperature (°C)	-33,0	-32,6	-27,9	-17,3	-6,4	4,1	11,0	9,7	2,3	-7,7	-20,3	-28,2	-12,2
Mean Monthly Precipitation (mm)	7,7	4,9	7,6	13,8	12,0	20,9	38,1	37,3	37,0	30,6	16,5	8,2	234,6

Temperature range	44,0 °C		Vegetation zone	Tundra (southern)
Extreme minimum	-50,6 °C		Permafrost type	Continuous

Table 19: Hall Beach (8 m), Northwest Territories (Canada)

Month	J	F	M	A	M	J	J	A	S	O	N	D	Year
Mean Monthly Temperature (°C)	-31,4	-32,7	-29,7	-20,9	-9,5	0,3	5,8	4,5	-0,5	-10,0	-20,8	-27,6	-14,4
Mean Monthly Precipitation (mm)	7,5	6,3	10,1	11,4	15,2	16,7	31,0	44,0	26,4	22,1	15,0	8,9	214,6

Temperature range	37,2 °C		Vegetation zone	Tundra (middle)
Extreme minimum	-54,1 °C		Permafrost type	Continuous

Hall Beach (Table 19), in the central parts of the Northwest Territories and at the northern end of the Canadian mainland, has the coldest extreme minimum temperature (-54,1 °C) within this section. It is situated in the central part of the Tundra vegetation-zone and has no mean monthly temperature above +10 °C. The total amount of precipitation is 214,6 mm - mostly available in late summer or "autumn" (see Figure 20 in the Appendix) - and the permafrost type is continuous permafrost.

The second most northern station within this section is Resolute Airport *(Table 20)* at a southern island of the Queen Elisabeth Islands (Northwest Territories). It has the lowest total amount of precipitation of all here chosen Central-Canadian station (131,4 mm) and can be located within the high tundra. In winter precipitation is very little and the mean monthly temperatures are around -30 °C. During the short summer the temperatures are only a few degrees above freezing. By that way you are having continuous permafrost. For this station see also *Figure 21* in the Appendix.

Table 20: Resolute Airport (67 m), Northwest Territories (Canada)

Month	J	F	M	A	M	J	J	A	S	O	N	D	Year
Mean Monthly Temperature (°C)	-32,1	-33,2	-31,4	-23,1	-10,9	-0,6	4,1	2,4	-5,1	-15,1	-24,5	-29,3	-16,5
Mean Monthly Precipitation (mm)	3,3	3,0	3,0	5,9	8,1	12,1	22,5	31,1	18,0	13,8	5,7	4,9	131,4

Temperature range	37,3 °C		Vegetation zone	Tundra (high)
Extreme minimum	-52,2 °C		Permafrost type	Continuous

Alert, nearly at the top of Ellesmere Island (Northwest Territories) and high up north in the Arctic, is the most northern station of the here shown Central-Canadian section. The data are given in *Table 21*. It is located in the high Tundra and within the zone of continuous permafrost. In the short summer, the temperatures hardly reaching above freezing, the total range is 37,3 °C. Compared to the station Resolute Airport, laying quite much south, you won't find that much differences. The climate-diagram for Alert is drawn in *Figure 22* in the Appendix.

Table 21: Alert (62m), Northwest Territories (Canada)

Month	J	F	M	A	M	J	J	A	S	O	N	D	Year
Mean Monthly Temperature (°C)	-32,1	-33,6	-33,2	-24,9	-11,7	-1,0	3,6	0,9	-10,2	-19,7	-26,6	-30,0	-18,2
Mean Monthly Precipitation (mm)	7,1	5,2	6,8	7,6	10,4	12,1	19,5	28,3	27,7	13,5	8,3	7,9	154,4

Temperature range	37,2 °C		Vegetation zone	Tundra (high)
Extreme minimum	-50,0 °C		Permafrost type	Continuous

Regarding the whole Central-Canadian section, you will see, that the temperature range increases going northwards only until Baker Lake (44, 0 °C) and than therefore beginning to decrease. Even the total amount of precipitation is getting smaller moving north, one hardly can say, that the most continental climate will be at high latitudes. Only in the most southern station of Medicine Hat you can find a little "dry-period" (see *Figure 13* in the Appendix). The Arctic tree line can be located at the station Ennadai Lake (mean July temperature 13,0°C) while the southern limit of discontinuous permafrost is found at La Ronge (mean annual temperature -0,5 °C) - there a (great) distance between these two points.

2.5. Alaska

As a last example, the most continental arctic climate-station of North America, Fort Yukon in northeastern Alaska (United States) is given in *Table 22*. Here the temperature range is 45 °C and the total amount of Precipitation only 172 mm. The precipitation is nearly available at the same level throughout the year, except in spring and December. For the fact, that the mean monthly temperature is above 10 °C during three month, you can locate this station within the northern parts of the Taiga and the zone of discontinuous permafrost. Comparing this stations with the above shown from the Siberian section, it is not that continental at all. In Fort Yukon the range is much lower and 70 mm more precipitation are available than in Verkhoyansk (see *Table 4*). The climate-diagram for the station Fort Yukon is given in *Figure 23* in the Appendix.

Table 22: Fort Yukon (127 m), Alaska (United States)

Month	J	F	M	A	M	J	J	A	S	O	N	D	Year
Mean Monthly Temperature (°C)	-30	-27	-17	-5	6	13	15	12	7	-4	-21	-29	-6,7
Mean Monthly Precipitation (mm)	13	10	8	6	12	18	22	30	16	15	14	8	172

Temperature range	45 °C		Vegetation zone	Taiga (northern)
Extreme minimum	-57,2 °C		Permafrost type	Discontinuous

As a result of this chosen examples for arctic climates one can say, that the most continental climate can be found in northern and central-eastern Siberia while the most oceanic climate may be found in Quebec (eastern parts of Canada).

The Arctic tree line and the southern limit or extension of discontinuous permafrost corresponds a lot with the existing climate. In oceanic climates, like Quebec, the two "lines" are nearly close together, while in continental climates there might be a big distance of about 1000 km between them, as it is in Siberia. But there will not only be differences in the localisation of these two important arctic "borders" but also differences in the measured temperatures. As you can see, in continental regions the Arctic tree line can be located at a mean July temperature of about 13 °C and the southern limit of discontinuous permafrost reaches not farther than regions with a mean annual temperature of around +1 °C. On the other hand in oceanic climates, the Arctic Tree line and the southern limit of discontinuous permafrost are recognised in colder regions (the mean July temperature is 10 °C respectively the mean annual temperature is -4,3 °C).

3. Permafrost and Arctic tree line

In the Introduction was already mentioned, that four types of permafrost could be distinguished by the extension (continuous, discontinuous, isolated an sporadic permafrost).

The largest permafrost regions can be found in central and eastern Siberia and in northern Canada and Alaska (see *Map 8* in the Appendix). At high latitudes in regions with a continental climate the permafrost may reach a high dept in the ground and covers large areas (zone of continuous permafrost). Further south, reaching the areas of discontinuous or isolated permafrost the dept is littler and you may find peat lands and isolated ice masses beside the permafrost areas. On the southern margin of permafrost, thick peats are found. (Harris, 1986) Overlaying permafrost is an "active" layer of rock or soil that will melt in summer and freezing again in winter (Oxford, 1978).

Decisive for the permafrost distribution are climatic and terrain factors like the ground temperature, the amount and duration of heat applied to the ground surface, the snow cover, the local relief and vegetation and hydrology (Harris, 1986). The snow cover, or strictly spoken the time of arrival of the snow cover, the snow dept and density and his duration, plays an important role on existence of permafrost, because it protects the underling ground from the cold air like an insulator. For example "the differences in the distribution of permafrost on the two sides of the Hudson Bay are due to the deeper snow cover in Quebec" (Harris, 1986, p.61). This fact is also shown in the climate tables of the Central-Canadian Section and the Quebec Section, which are given in chapter 2. The topography or relief influences the microclimate and the amount of solar radiation received by the ground surface. It also will influence the snow cover.

In my opinion the most important factor for permafrost beside snow cover is the hydrology. "Moving water is an effective thawing and erosive agent." (Harris, 1986) It can penetrate into large areas and drainage them - therefore the permeability is most deciduous. If lakes are not freezing to the bottom (large) taliks can be found (Harris, 1986).

Trees are shading the ground from solar radiation in summer and acting as radiators of heat in winter (Harris, 1986). By that way they are influencing the distribution of permafrost, but the vegetation and also the occurrence of trees are as well influenced by the existence of permafrost and therefore the dept of the active layer. But the permafrost limit does not correspond with the Arctic tree line at all. While in eastern Canada, the Arctic tree line and the limit of discontinuous permafrost correlate in a large way with its position, in Siberia the permafrost reaches far into the Taiga-Forest (Sveinbjörnsson, 1992).

The Arctic tree line, or also known as Circumpolar or latitudinal tree line, is the "border" between the two vegetation zones Taiga and Tundra. Another common word for it is Arctic timberline. "The sharpness and the regularity of the Taiga-Tundra boundary reflect both the topography and the history of the land surface." (Sveinbjörnsson, 1992, p.239) Its location is to a large extent determined by the climate. The closely interrelated factors solar radiation and temperature are the most important ones for the circumpolar tree line (Sveinbjörnsson, 1992). The growing of trees thereby can be related to the air temperature - you will need a minimum of 30 days with a mean daily temperature above 10 °C. Köppen, a German biologist and geographer, already correlated in the thirties of last century the tree line with the mean temperature of 10 °C of the warmest month (Sveinbjörnsson, 1992) - the 10°C isotherm. The soil temperature may also play an not unimportant roll as also the dept and duration of snow cover for the survival of trees.

However the Arctic latitudinal tree line cannot easily marked. It is a much broader zone of transition than the altitudinal timberline (Young, 1989). A reason for that fact is, that, as you can see in the above given climate-tables, the temperature will change only "less per unit of latitudinal distance" (Young, 1989, p.154). In the zone north of the heavy boreal forest, you will reach the forest Tundra. Here you can often find little forests or some lone-standing or grouped tress, even if you are already many kilometres behind the so-called tree line. After the tall-standing trees are fully gone, you may find dwarf shrubs, Krummholz and other forms of growing of tree-species, e.g. birches.
The location and the structure of the timberline are often heavily influenced by human activity (Sveinbjörnsson, 1992). Over and above that, the relief is an important factor for the growing tress. In some against wind and cold temperatures protected areas, like river valleys, you might even find trees or a small forest high up north of the so-called tree line. Such phenomena can be for example found at some few micro sites of the Mackenzie Valley in northern Canada (Sveinbjörnsson, 1992) or high arctic Canadian islands.

4. Conclusions:

The Arctic tree line and the southern limit of discontinuous permafrost are depending in a large way on the existing climate, but this is not the only limited factor for these important Arctic borderlines. The relief or landscape is also, beside others, a much important factor, e.g. the location of the mountains. These two so-called boarders will even sometimes influence each other. They are dividing the Sub arctic from the Arctic region.

Nevertheless, the climate - the temperature and the precipitation - is in my opinion the most important causing factor. Regarding the maps of the vegetation-zones of Eurasia (see *Map 6* in the Appendix) and North America (see *Map 7* in the Appendix) one can see, that the Boreal Forest-zone, called Taiga, reaches too much higher latitude in continental climates than in oceanic regions. This fact is clearly to be seen in Central Canada. It is also obvious, that distribution of permafrost must have something in common with the amount of precipitation and the snow cover. A minimum of 30 days with a mean daily temperature above 10 °C will be needed for a successful trees growth. The 10 °C isotherm however cannot be used exactly for all arctic climate regions - in continental regions the mean July temperature must be a few degrees higher.

The in chapter 2 discussed four south-north sections are showing the wide range of climates that can be found within the Arctic and Sub arctic. For the reason, that the above given climate data are averages values of a longer period - normally 30 years - they are not showing the variability, that can arise from one year to the next in an area. Here it is mean, that some winters can be extremely cold while other a relatively warm or that the amount of fallen snow in winter can be much more in some years than in the average.

5. References:

Canadian Climate Program: Canadian Climate Normals 1961-1990, vol. 2 - Prairie Provinces, pub. by the Canadian Ministry of Supply and Service, Ottawa 1993

Canadian Climate Program: Canadian Climate Normals 1961-1990, vol. 3 - The North. Yukon Territory and Northwest Territory, pub. by the Canadian Ministry of Supply and Service, Ottawa 1993

Canadian Climate Program: Canadian Climate Normals 1961-1990, vol. 6 - Quebec, pub. by the Canadian Ministry of Supply and Service, Ottawa 1993

Harris, S. A.: The permafrost environment, London/Sydney 1986

Oxford Economic Atlas of the Word, 4th ed., Oxford 1978, p.1

Sveinbjörnsson, B.: Arctic Tree Line in a Changing Climate, in: Chapin III, F. S., Jefferies, R.L., Reynolds, J.F., Shaver, G.R., Svoboda, J and Chu, E. W. (Eds.): Arctic Ecosystems in a Changing Climate. An Ecophysiological Perspective, San Diego, 1992, pp. 239-256

Young, S.B.: Gateway to the Arctic. The northern Forest and the Timberline, in: S.B. Young: To the Arctic. An introduction to the far northern world, New York 1989, pp. 151-181

6. Appendix:

Part 1 - Climate Diagrams

I. Siberian Section

Figure 2:

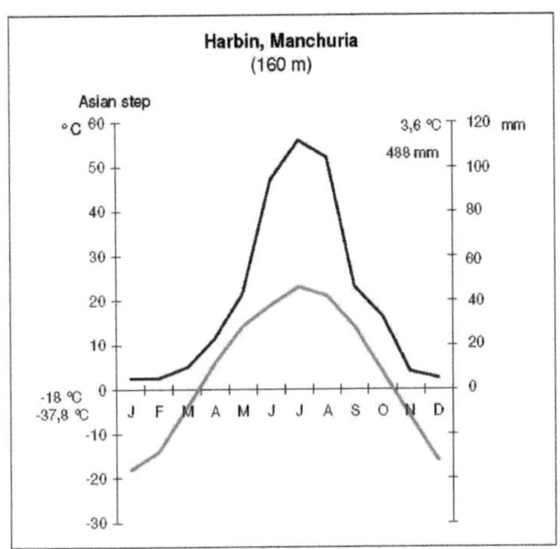

Figure 3:

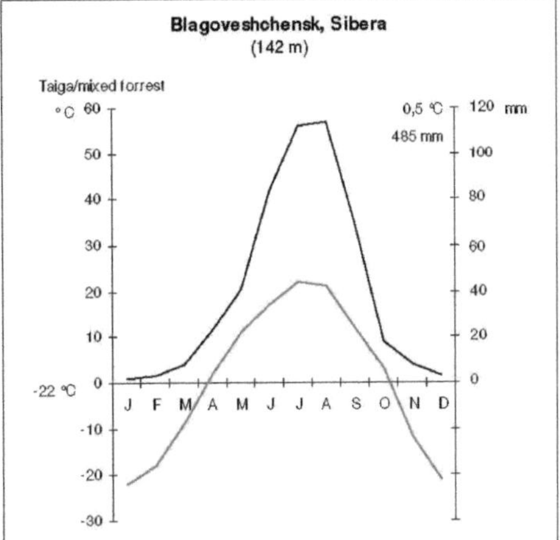

Figure 4:

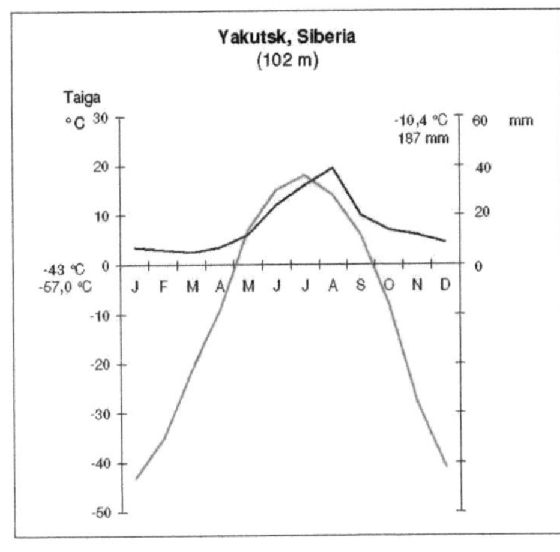

Figure 5:

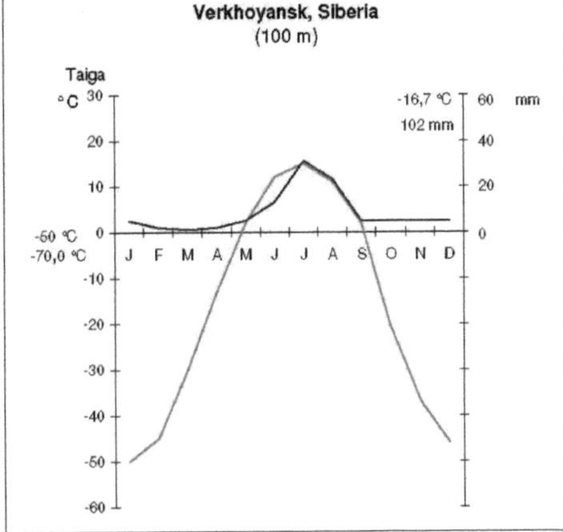

Figure 6:

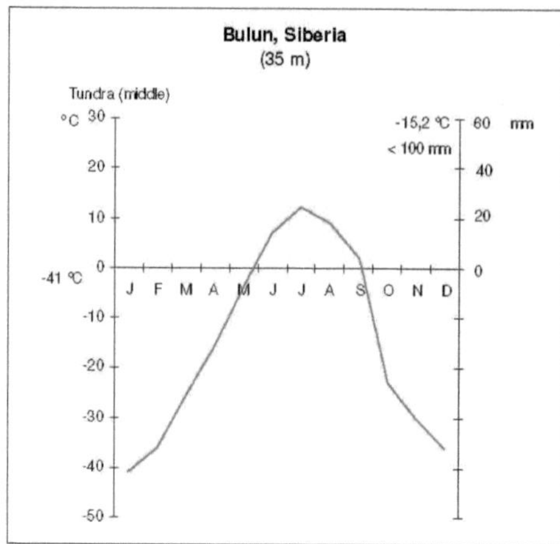

II. Komi-Nanette Section

Figure 7:

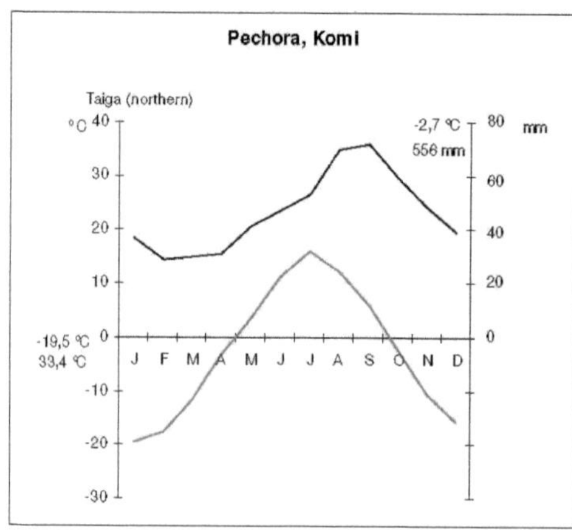

Figure 8:

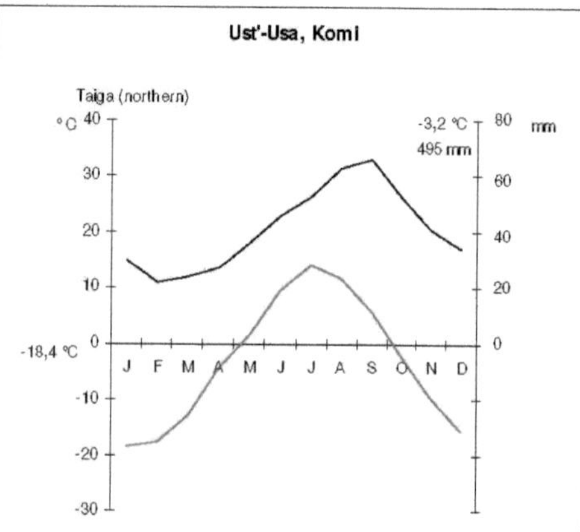

Figure 9:

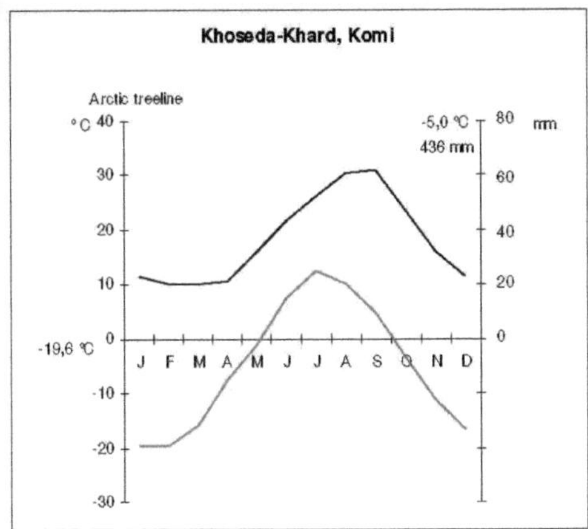

III. Quebec Section

Figure 10:

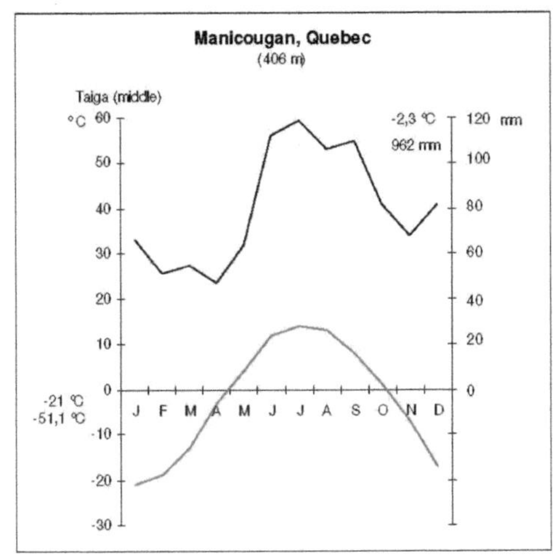

Figure 11:

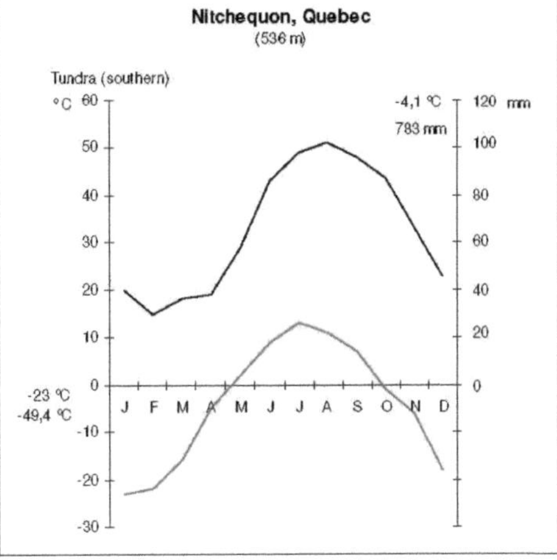

Figure 12:

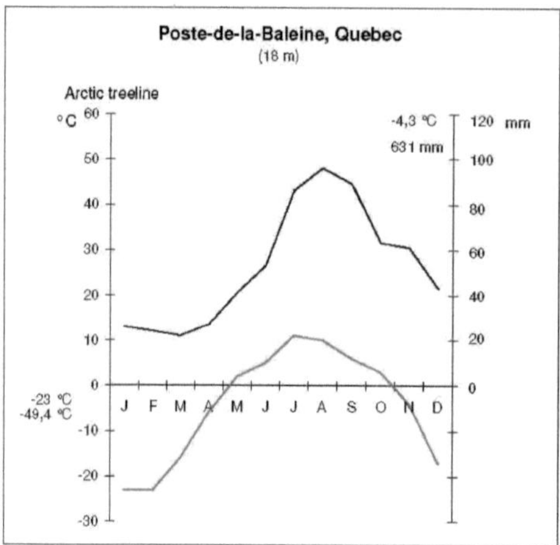

IV. Central-Canada Section

Figure 13:

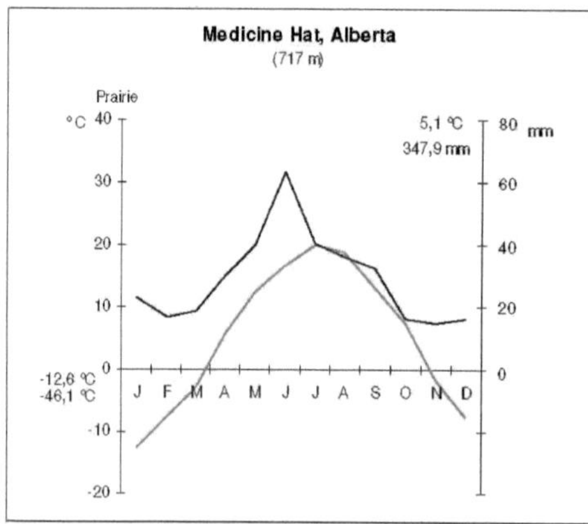

Figure 14:

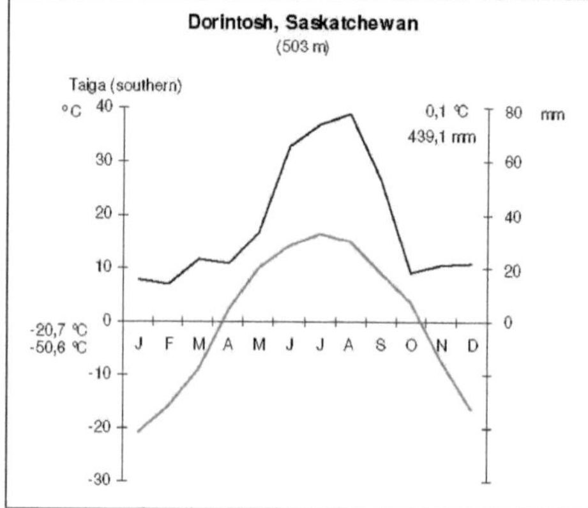

17

Figure 15:

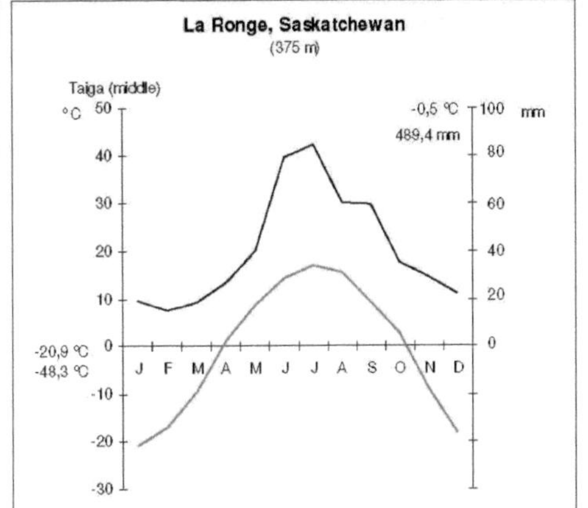

Figure 16:

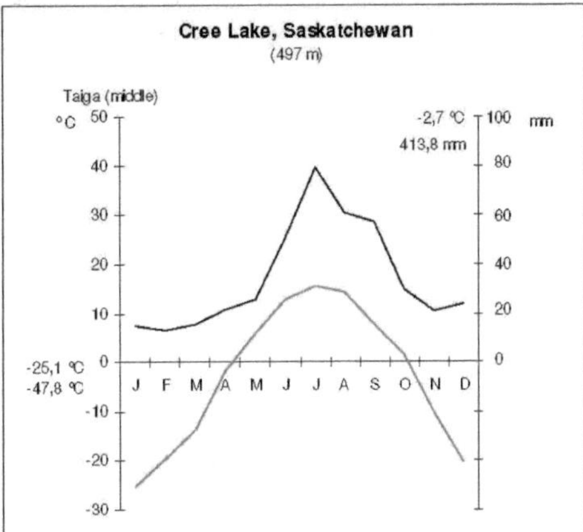

Figure 17:

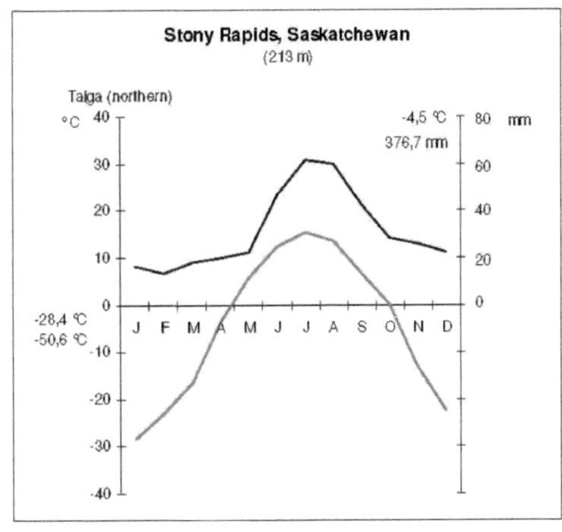

Figure 18:

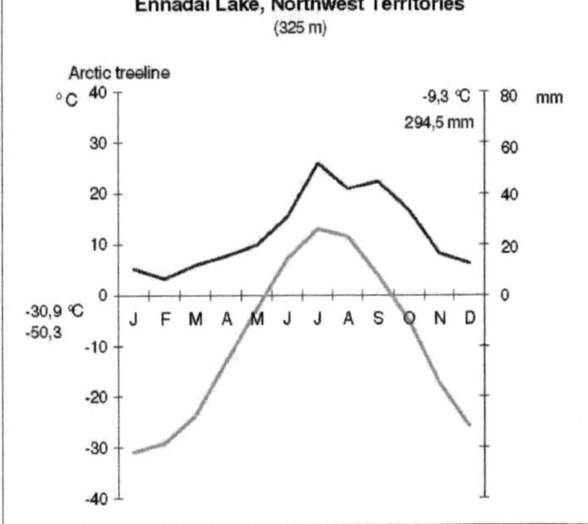

Figure 19:

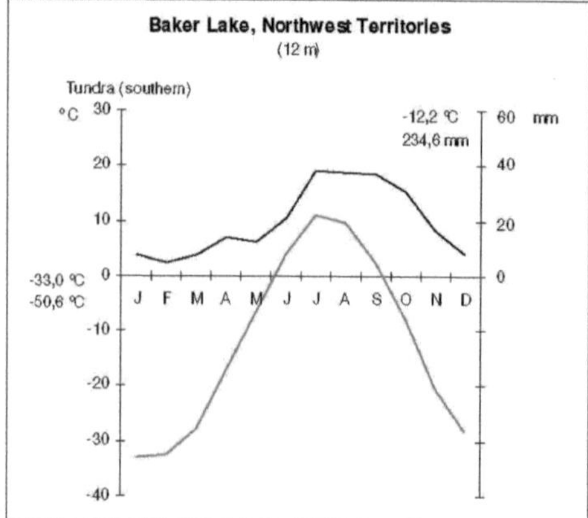

Figure 20:

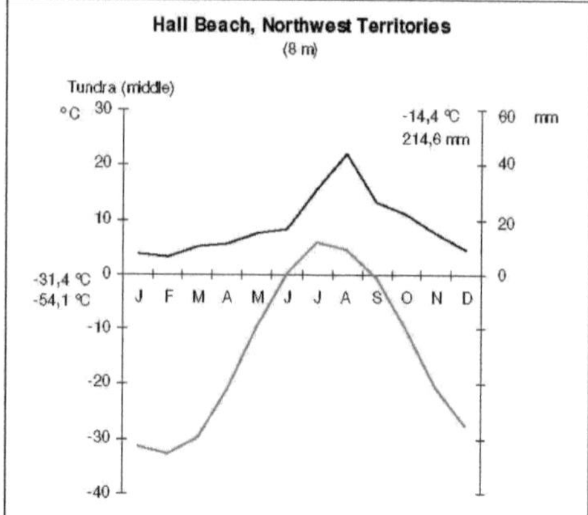

Figure 21:

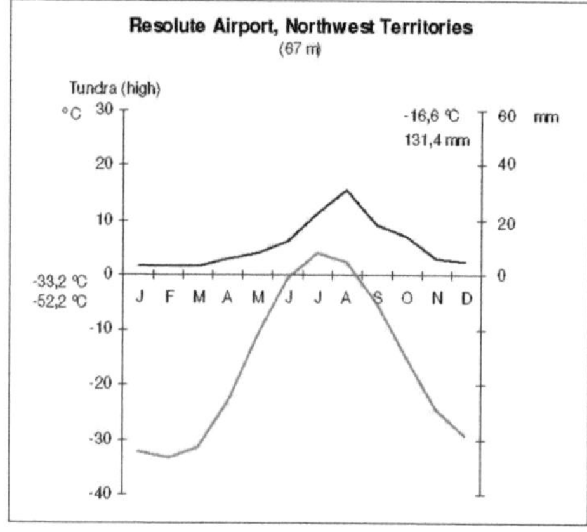

Figure 22:

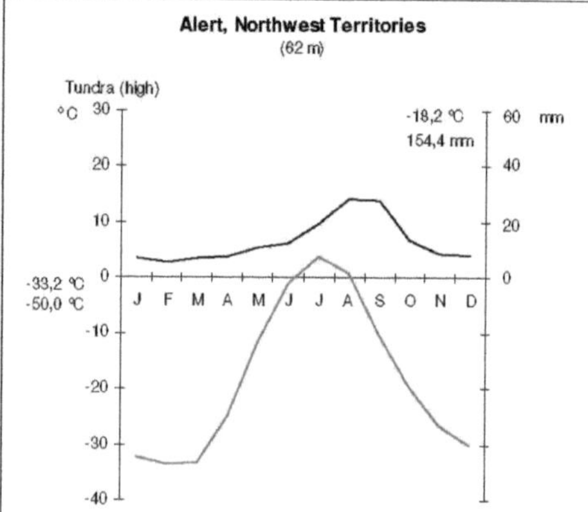

V. Alaska

Figure 23:

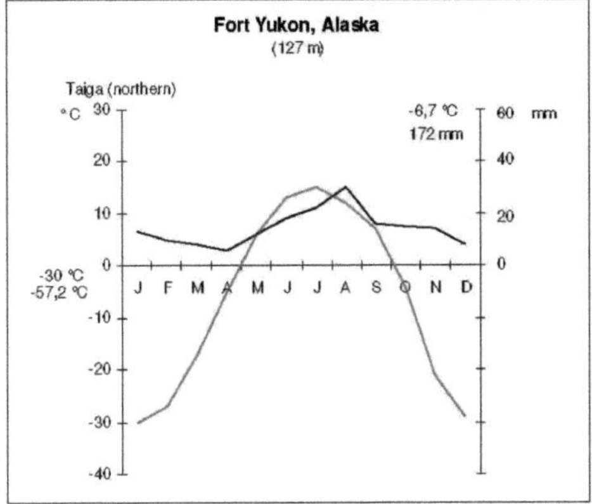

Part 2 - Maps

Map 1: Siberian Section

Source: *The Tines Atlas of the Word. Concise edition, 6ᵗʰ ed., London 1992, p.51*

Map 2: Komi Section

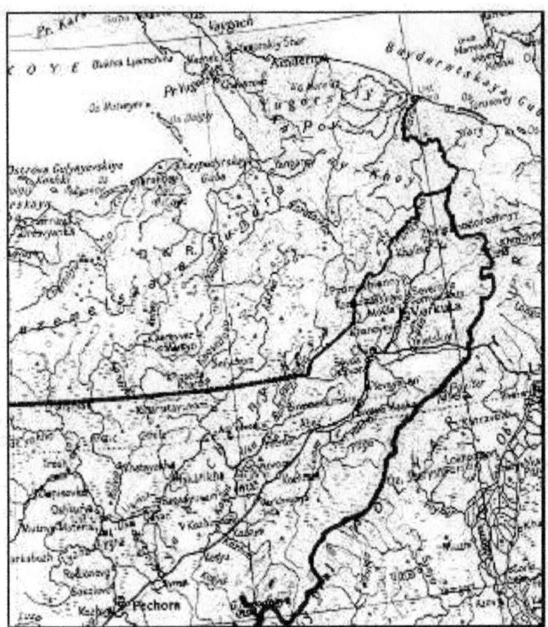

Source: *The Times Atlas of the World. Comprehensive edition*, 9th ed., London 1992, p.47

Map 3: Quebec Section

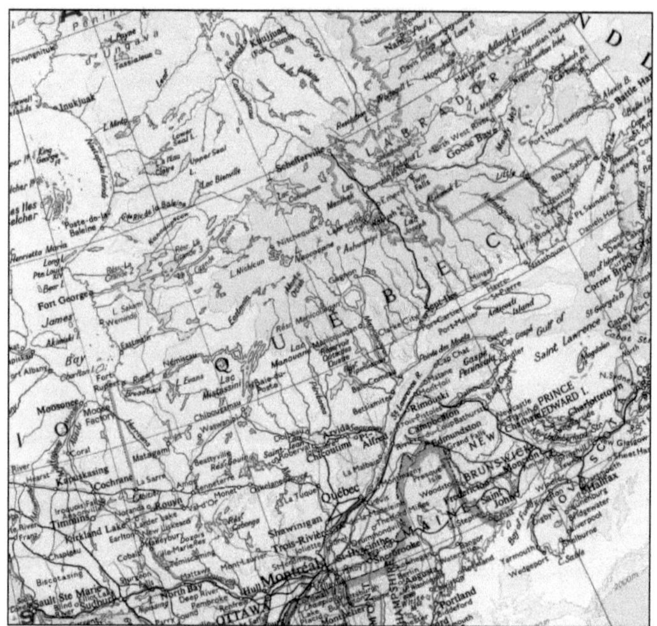

Source: The Tines Atlas of the Word. Concise edition, 6th ed., London 1992, p.115

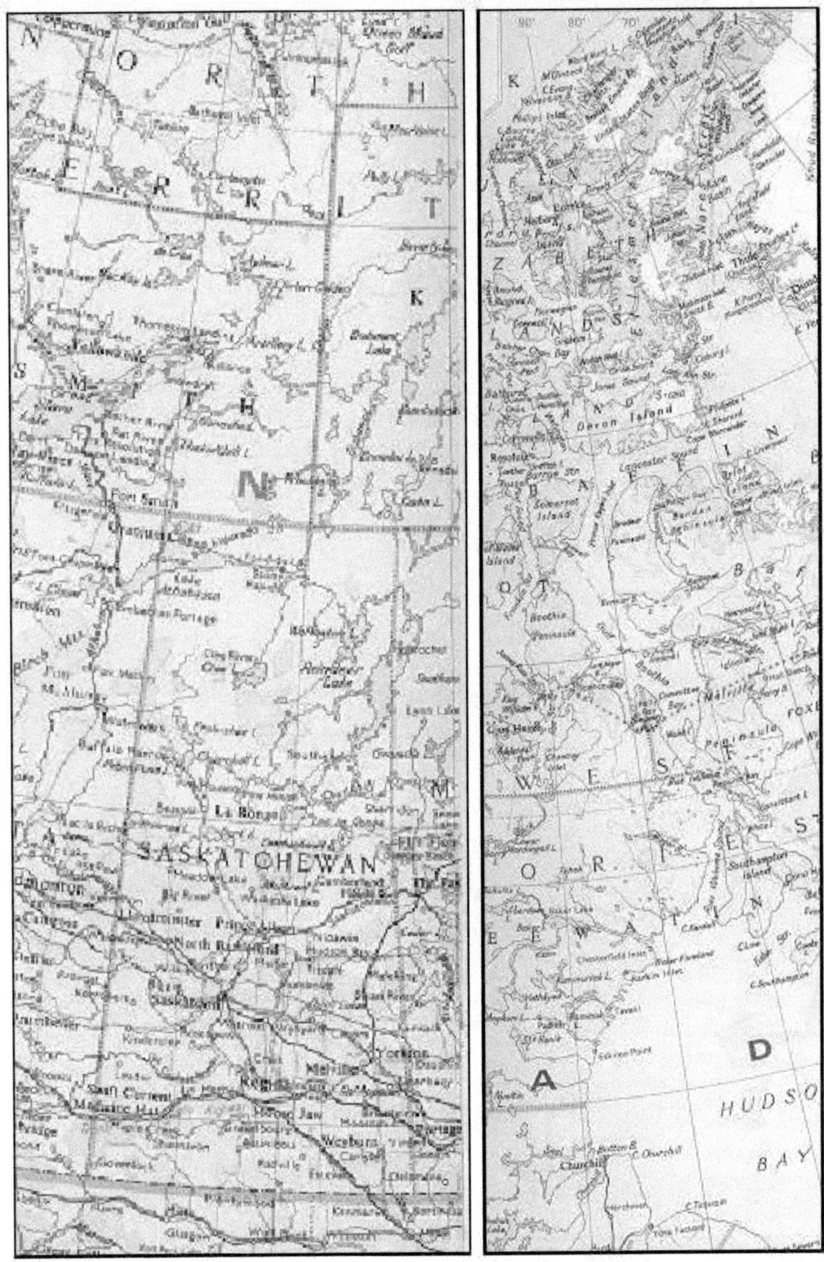

Source: *The Times Atlas of the Word. Concise edition. 8th ed. London 1992, p.114/115*

Map 5: Alaska

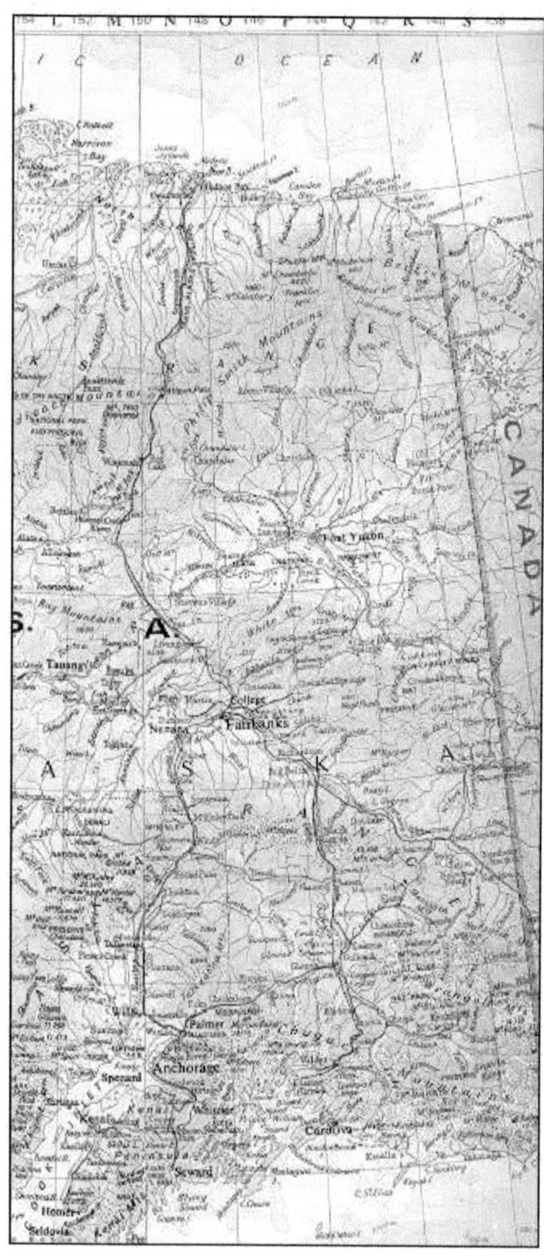

Source: *The Times Atlas of the Word. Concise edition, 6* ed., London 1992, p.116*

Map 6: Vegetation-zones of Eurasia

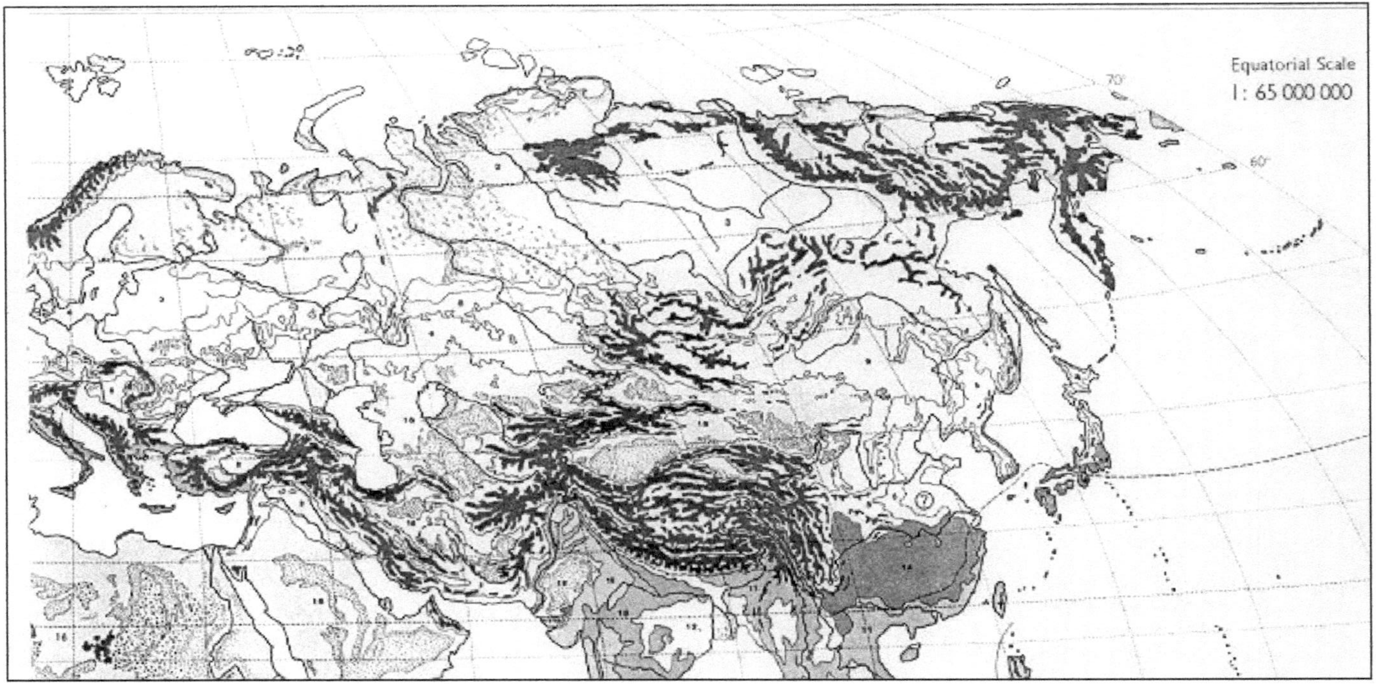

Source: The Times Atlas of the Word. Comprehensive edition, 9ᵗʰ ed., London 1992, p.5

[Explanations see after Map 7 on next page]

Map 7: Vegetation-zones of North-America

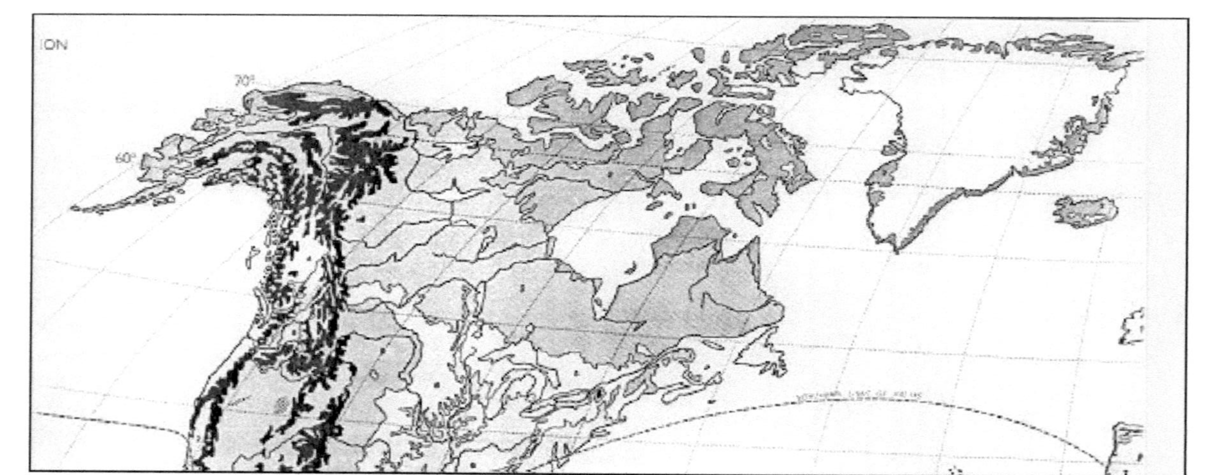

Source: The Times Atlas of the Word. Comprehensive edition, 9th ed., London 1992, p.5

Legend:

number	vegetation Zone	color (if marked)	number	vegetation Zone	color (if marked)
1	Mountain Vegetation	black	10	Savannah (grass and shrub)	
2	Tundra (moss, lichen, etc.)	dark gray	11	Tropical Rain Forest ("selva")	
3	Borreal Forest ("Taiga")	light gray	12	Monsoon Forest (moist decidious)	
4	Conifer Forest (pine. Spruce and larch)		13	Dry Tropical Forest (semi-decidious)	
5	Mix-Forrests, Mid-Latitudes (boradleaf and conifer)	dark green	14	Sub Tropical Forest (Dry and wet broadleaf evergreen)	
6	Broadleaf Forest (decidious)	light green	15	Dry Tropical Scrub Thorn Forest	
7	Mediteranian Scrub (citrus, olive, agave, etc.)		16	Dessert Vegetation (xerophytic shrub, grass and catus)	
8	Prairie (long grass)	orange			
9	Steppe (short grass)	yellow	The Arctic tree line is marked red		

Map 8: The distribution of permafrost in the Arctic

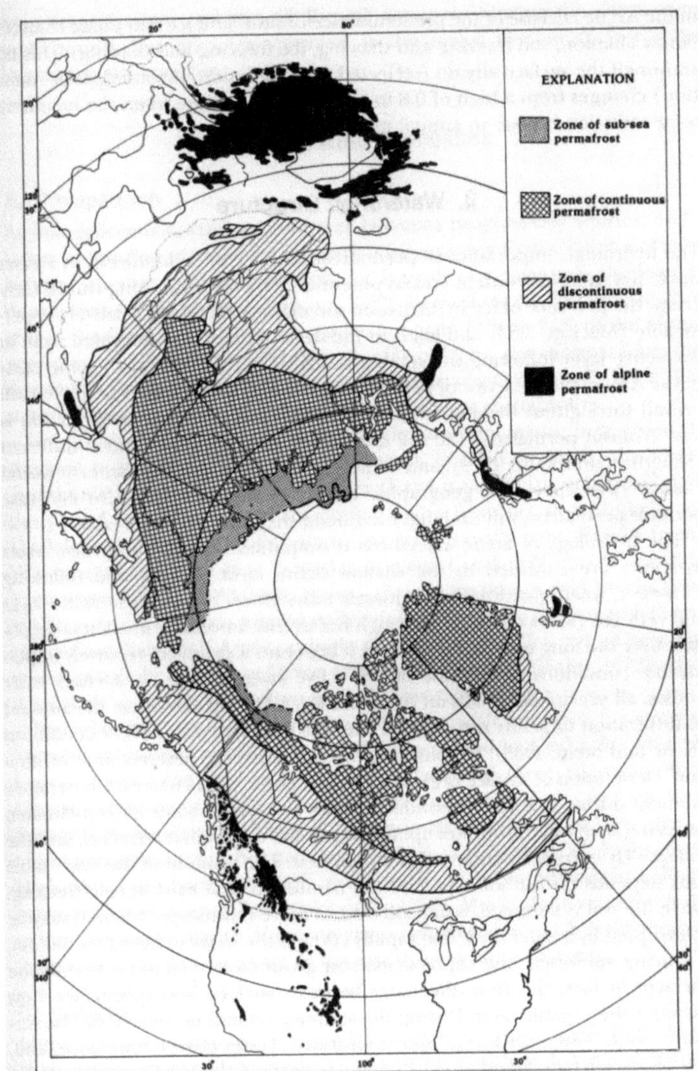

Figure 1 Distribution of permafrost in the Northern Hemisphere. Some isolated areas of alpine permafrost lie off the map; these, and similar areas within the map limits are not shown. [Reprinted from Péwé (1983) with permission from the Institute of Arctic and Alpine Research.]

Source: F. Stuart Chapin III et al.: Arctic Ecosystems in a Changing Climate. A changing Eco-physiological Perspective, San Diego 199, p.37